Leandro Heleno Guimarães Lacerda

Resistant Hypertension: Potential Biochemical Markers

Leandro Heleno Guimarães Lacerda

Resistant Hypertension: Potential Biochemical Markers

Participation in antihypertensive control

Imprint

Any brand names and product names mentioned in this book are subject to trademark, brand or patent protection and are trademarks or registered trademarks of their respective holders. The use of brand names, product names, common names, trade names, product descriptions etc. even without a particular marking in this work is in no way to be construed to mean that such names may be regarded as unrestricted in respect of trademark and brand protection legislation and could thus be used by anyone.

Cover image: www.ingimage.com

This book is a translation from the original published under ISBN 978-613-9-67776-4.

Publisher:
Sciencia Scripts
is a trademark of
Dodo Books Indian Ocean Ltd. and OmniScriptum S.R.L publishing group

120 High Road, East Finchley, London, N2 9ED, United Kingdom
Str. Armeneasca 28/1, office 1, Chisinau MD-2012, Republic of Moldova, Europe
Printed at: see last page
ISBN: 978-620-8-17924-3

Copyright © Leandro Heleno Guimarães Lacerda
Copyright © 2024 Dodo Books Indian Ocean Ltd. and OmniScriptum S.R.L publishing group

SUMMARY

ACKNOWLEDGMENTS

To my advisor, Dr. Valéria Cristina Sandrim, for believing in my potential, for her patience and support, which resulted in high-quality work.

To Dr. Karla Fernandes and Dr. Vanessa Andrade for their constant support and willingness to guide me through crucial stages of this project.

To my mother, Patricia dos Santos Guimaraes, for her love, affection, attention, for giving me the opportunity to study and for her efforts, encouragement and constant support, which were crucial to the completion of this work and the success of another stage in my life.

To my family, who have always believed in my potential and made themselves available whenever I needed support and affection.

To all those who participated in some stage of this study.

To God and the master Jesus for giving us health and wisdom and for giving us the gift of life and the opportunity to experience it.

SUMMARY

Introduction: Systemic arterial hypertension (SAH) stands out as the biggest independent risk factor related to cardiovascular disease (CVD) and remains the biggest modifiable risk factor, despite significant advances in knowledge of its pathophysiology and the availability of effective methods for its treatment. There are approximately 18 million people in Brazil with hypertension and only 30% of them have the disease under control, which can increase the risk of health problems in this population. Resistant arterial hypertension (RAH) affects an average of 30% of the adult population, around 1.2 billion people worldwide. Although the exact prevalence of ARH has not yet been established, it is estimated that this condition affects 12-15% of hypertensive patients. Matrix metalloproteinase 9 (MMP-9) is involved in the degradation of extracellular matrix (ECM) components and is mainly inhibited by its natural endogenous inhibitor, tissue inhibitor of matrix metalloproteinase 1 (TIMP-1), both of which participate in the vascular remodeling process and may influence vascular stiffness. Objective: The aim of this study was to compare plasma levels of MMP- 9, TIMP-1 and the MMP-9/TIMP-1 ratio between controlled resistant hypertensive patients (CRHT) and uncontrolled resistant hypertensive patients (NCRHT) and subsequently to verify the correlation between plasma levels of these biomarkers and clinical variables (gender, age, BMI, SBP, DBP, PP, OPV, total cholesterol, LDL, HDL, triglycerides, urea, creatinine, glycemia, uric acid, aldosterone and renin) in these groups. Materials and Methods: A total of 76 individuals diagnosed with ARH who were being followed up at the Resistant Hypertension Outpatient Clinic at the University of Campinas - UNICAMP/SP took part in this study. Of the total number of patients, 35 were classified as HTRNC and 41 were classified as HTRC. The patients underwent dosing of plasma MMP-9 and TIMP-1, and arterial stiffness was assessed using the pulse wave velocity (PWV) method. Results: The clinical variables age, DBP, glycemia and uric acid correlated significantly with TIMP-1 and the variables BMI and renin correlated significantly with MMP-9. The other clinical variables did not show statistically significant correlations with the biomarkers in question. Conclusion: A greater association between TIMP-1 and uncontrolled resistant hypertension was evidenced in this study, suggesting that TIMP-1 plays an important role in the mechanisms of development and control of ARH.

Keywords: Resistant hypertension, MMP-9, TIMP-1, vascular stiffness.

1 INTRODUCTION

1.1 Hypertension

1.1.1 Concept, diagnosis and prevalence.

Systemic arterial hypertension (SAH) is a multifactorial clinical condition characterized by high and sustained levels of blood pressure (BP), which is diagnosed by casual measurement. It is often associated with functional and/or structural alterations in target organs (heart, brain, kidneys and blood vessels) and metabolic alterations, with a consequent increase in the risk of fatal and non-fatal cardiovascular events (VI BRAZILIAN guidelines ON HYPERTENSION, 2010).

According to the VI Brazilian Guidelines for Hypertension (2010), hypertensive adults are those whose Systemic Blood Pressure (SBP) reaches 140 mmHg or more, and/or whose Diastolic Blood Pressure (DBP) is equal to or greater than 90 mmHg, on two or more occasions, in the absence of antihypertensive medication (table 1).

Table 1. Classification of blood pressure according to casual measurement in the doctor's office (> 18 years).

Classification	Systemic pressure (nınıHg}	Diastolic pressure InimHg)
Qtima	< 120	<80
Normal	< 130	<85
Lrnftrofe*	130-139	B5-B9
Stage 1 hypertension	140-159	90-99
Stage 2 hypertension	160-179	100-109
Stage 3 hypertension	≥ 180	≥ 110
Isolated Systemic Hypertension	≥ 140	<90

When systemic and diastolic pressures are in different categories, the higher should be used to classify blood pressure.

*-P⅛ njje⅛ *normal-high or pre-hypertension sSa ⅛ft!|oɪ que so ßguivalem* na *Iitevaiura.*

According to Andrade and collaborators (2006), it is defined as a clinical state in which the individual presents repeated imbalances in BP values that confer a significant increase in the risk of cardiovascular events, in the short and long term, justifying a therapeutic program. It is a chronic disease, usually asymptomatic, and must therefore be treated and controlled for the rest of one's life. It has a high medical and social cost, as it is one of the most important risk factors for the development of cardiovascular diseases (RENOVATO;

TRINDADE, 2004).

Arterial hypertension (AH) can be clinically classified in two ways: primary, systemic or essential arterial hypertension (SAH, AH or hypertension), which is an isolated manifestation of an unknown cause, and secondary hypertension, which is a manifestation resulting from other diseases such as nephropathies, aortic coarctation and pregnancy toxemia, among other causes (ANDRADE *et al.*, 2006).

The term arterial hypertension, commonly referred to as hypertension or high blood pressure, can be caused by loss of elasticity in the arteries, cardiovascular and renal complications, which cause BP to rise above 140 x 90 mmHg (CECIL, 2001). Cardiac output and peripheral vascular resistance are the determinants of BP and any alteration in one or the other, or both, interferes with the maintenance of normal pressure levels (KRIEGER *et al.*, 1999).

SAH has a high prevalence and low control rates. It is considered one of the main modifiable risk factors (RF) and one of the most important public health problems. Mortality from cardiovascular disease (CVD) increases progressively with a linear, continuous and independent rise in BP above 115/75 mmHg. In 2001, around 7.6 million deaths in the world were attributed to high BP, 54% due to strokes and 47% to ischemic heart disease (IHD), the majority in low and medium economic development countries and more than half in individuals aged between 45 and 69. In Brazil, CVD has been the leading cause of death and is responsible for a high frequency of hospitalizations, leading to high medical and socioeconomic costs (BRAZILIAN GUIDELINES ON HYPERTENSION, 2010).

According to the VI Brazilian Guidelines for Hypertension (2010), population surveys in Brazilian cities over the last 20 years have shown a prevalence of hypertension of over 30%. Considering BP values ≥ 140/90 mmHg, 22 studies found prevalence rates between 22.3% and 43.9% (average 32.5%), with more than 50% between 60 and 69 years old and 75% over 70 years old. Between genders, the prevalence is 35.8% in men and 30% in women, similar to other countries. A quantitative systematic review from 2003 to 2008 of 44 studies in 35 countries revealed an overall prevalence of 37.8% in men and 32.1% in women.

1.1.2 Factors Associated with High Blood Pressure

Hypertension depends on the interaction between genetic predisposition, environmental factors (such as psycho-emotional stress and salt intake, which reinforce and precipitate the hemodynamic expression of polygenic predisposition) and early structural adaptations of the heart and vessels, such as pathological vascular remodeling and cardiac hypertrophy

(MANO, 2003).

Although it is not yet completely known how these interactions occur, it is known that AH is accompanied by functional alterations in the sympathetic autonomic nervous system, the kidneys, the renin angiotensin system, as well as other humoral mechanisms and endothelial dysfunction. Thus, hypertension is the result of various structural alterations in the cardiovascular system that both amplify the hypertensive stimulus and cause cardiovascular damage (MANO, 2003), and is considered a syndrome because it is often associated with a cluster of metabolic disorders, such as obesity, increased insulin resistance, diabetes mellitus and dyslipidemia, among others (ROSARIO, 2009).

It is extremely important to address the non-modifiable and modifiable risk factors with regard to hypertension, as these are part of the patient's lifestyle and can be identified in that lifestyle. Non-modifiable risk factors include age, heredity and gender (VI BRAZILIAN GUIDELINES ON HYPERTENSION, 2010).

The increase in blood pressure with advancing age has been observed in several studies, although according to the III Brazilian Consensus on Hypertension (2001), this increase does not represent normal physiological behavior (siLVA; sOUZA, 2004).

According to Barreto-Filho and Krieger (2003), one third of the factors involved in the pathophysiology of hypertension can be attributed to genetic factors. They cite the BP regulatory system and salt sensitivity as examples. The authors make it clear that AH can be understood as a multifactorial syndrome with a poorly elucidated pathogenesis, in which complex interactions between genetic and environmental factors cause a sustained rise in blood pressure. In approximately 90% to 95% of cases there is no known etiology or cure, and BP control is achieved through lifestyle changes and pharmacological treatment.

In order to make a therapeutic decision, it is necessary to stratify overall cardiovascular risk, which will take into account, in addition to BP values, the presence of additional risk factors, target organ damage and cardiovascular diseases (Vi DiRETRiZEs BRAsiLEiRAs DE HiPERTENsăo ARTERiAL, 2010).

Table 2 shows the risk stratification associated with patients with hypertension, comparing normotensive and hypertensive individuals.

Table 2. Risk stratification of hypertensive patients.

Other risk factors or diseases	Norniotension			Hypertension		
	Optimal SBP < 120 ou DBP <80	Normal SBP 120-129 or PAD 80-84	Borderline SBP 130-139 or DBP 85-89	Stage 1 PAS140-159 PAD 90-99	Stage 2 SBP160-179 DBP 100-109	Stage 3 SBP >180 DBP >110
Neiiliumfatorderisco	Riscobasal	Riscobasal	Riscobasal	Additional low risk	Moderate additional risk	Cyclonal high risk
1 to 2 risk factors	Additional low risk	Additional low risk	Additional low risk	Moderatoradditional risk	Moderate additional risk	Very high cyclonal risk
> 3 risk factors, LOA or SM-DM	Moderatoradditional risk	Moderate additional risk	Traditional high risk	Traditional high risk	Altonscoadlcional	Very high cyclonal risk
Associated clinical conditions	Very high additional risk	Very high additional risk	Very high additional risk	Very high additional risk	Very high performance	Very high cyclonal risk

Dii: diabetes: melilo: LOA: lithium organ-aho: DBP: diabolic arterial pressure; SBP: arterial pressure :i:tolica: Metabolic syndrome.

Modifiable risk factors include social habits such as alcohol consumption, smoking, sedentary lifestyles and dietary patterns. The III Brazilian Consensus on Hypertension (2001) provides information on non-drug treatment, correlating it with lifestyle changes to · prevent and/or help treat hypertension. Excess body weight, increased salt/sodium intake, excessive consumption of alcoholic beverages, smoking and a sedentary lifestyle are proven habits that raise BP. Modifying these factors is related to a reduction in BP and consequently to a better quality of life for the patient. In addition to these factors, increasing potassium intake (which has an antihypertensive effect and protects against cardiovascular damage), controlling morbidities such as dyslipidemia and diabetes mellltus, adopting anti-stress measures and avoiding drugs that can raise BP are all measures to combat the onset of hypertension.

It is clear that almost all non-drug measures depend on permanent lifestyle changes. Because the approach to hypertension is aimed at different objectives, medical action benefits from a multi-professional approach. It is worth emphasizing that the involvement of the hypertensive patient's family members in the pursuit of the goals to be achieved by lifestyle modifications is of fundamental importance (III BRAZILIAN CONSENSUS OF ARTERIAL HYPERTENSION, 2001).

1.2 Resistant or refractory hypertension

In some hypertensive patients we can observe a maintenance of high blood pressure levels even with the use of antihypertensive polytherapy.

Currently, resistant arterial hypertension (RAH) is defined as the maintenance of high blood pressure levels (> 140 x 90 mmHg) despite the use of at least three antihypertensive drugs in optimized doses from different pharmacological classes, one of which is a diuretic, or that controlled with the use of four or more antihypertensive drugs. This entity is becoming increasingly common in daily clinical practice, and its greater potential for target organ damage makes accurate etiological diagnosis and early pressure control essential (MARTINS et al., 2010).

Despite the appropriate prescription of three or more antihypertensive agents and appropriate follow-up in specialized AH treatment centers, a significant number of hypertensive patients (approximately 40%) do not achieve these goals and are considered refractory (SOUZA, 2009). Although its real prevalence is unknown, it is known that it occurs in approximately 40% of patients with hypertension and is more commonly found in elderly patients, becoming increasingly common in daily clinical practice, and due to its greater potential for target organ damage, accurate etiological diagnosis and early pressure control are essential (SCUOTTO et al., 2009; MARTINS et al., 2010).

Furthermore, in order to characterize a state of refractoriness to therapy, it is necessary to be certain of the degree of adherence and responsiveness to treatment (FAVARELLI, 2001).

Patients who are intolerant of diuretics and have uncontrolled blood pressure under a three-drug regimen from other classes should also be considered resistant hypertensive patients (RHT) (CALHOUN et al., 2008).

For a better understanding of the current study, it is necessary to define uncontrolled resistant hypertension (uncontrolled resistant hypertension). Uncontrolled resistant hypertensives are those individuals who, even with the use of four or more antihypertensive drugs, one of which is a diuretic, do not reach ideal blood pressure levels (< 140 x 90 mmHg), characterizing a lack of responsiveness to therapy.

Resistance to antihypertensive therapy is usually multifactorial. However, pseudo-resistance, contributing factors and secondary hypertension are factors that play an important role in characterizing ARH (SOUZA, 2009).

1.2.1 Causes of resistance

1.2.1.1   Pseudo-resistance

Resistance, which acts on responsiveness to antihypertensive polytherapy, is usually influenced by various factors. It is important to ensure that blood pressure is measured correctly in order to make a more accurate diagnosis, thus providing more appropriate treatment and, consequently, successful pharmacological therapy for patients. In addition to the correct measurement of blood pressure, the correct choice of the type of antihypertensive drug must be made, as well as their respective optimized doses and an accurate assessment of adherence and responsiveness to treatment, which can directly influence the success of blood pressure control, i.e. the goal of the stipulated pharmacological therapy.

Scuotto and colleagues (2009) report that the first step in investigating ARH is to check for pseudo-refractoriness. To this end, the correct BP measurement technique should be observed, adherence to drug and non-drug therapy should be checked, the white coat effect should be diagnosed, and lifestyle changes should be made with weight reduction, a low-sodium diet, regular physical activity, and smoking and alcohol cessation.

In this context, we can say that pseudo-resistance occurs due to failures to check blood pressure, incorrect pharmacological therapy (doses and drugs) being prescribed to a particular patient, as well as non-adherence to the prescribed therapy.

1.2.1.2   Adherence to treatment

Diagnosing AH is not enough. What is essential is to conduct the treatment correctly and convince the patient of the need to adhere to and control the disease in order to improve their quality of life without deteriorating their condition (LESSA, 2006).

The term "adherence to treatment" can be understood as the way in which the patient interprets the doctor's instructions regarding the use of medication, dietary fidelity and making changes to lifestyle habits, as well as following the prescription fully and correctly, taking into account relevant items such as dosage, dose and schedule, i.e. how the patient obeys the doctor's instructions regarding the stipulated therapy.

We can cite the following as determinants of non-adherence to anti-hypertensive treatment: **1.** Lack of knowledge on the part of the patient about the disease or motivation to treat an asymptomatic and chronic disease; **2.** Low socio-economic level, cultural aspects and mistaken beliefs acquired from experiences with the disease in the family context, and low self-esteem; **3.** Inadequate relationship with the health team; **4.** Long waiting times, difficulty in making appointments, lack of contact with absentees and those who leave the service; **5.** High cost of medication and occurrence of undesirable effects; **6.** Interference in quality of life after starting treatment (VI BRAZILIAN GUIDELINES ON ARTERIAL HYPERTENSION, 2010).

Hypertension is controlled through the active participation of the hypertensive patient and the co-participation of the family, health professionals and the correct performance of health programs managed by institutions of any kind (LESSA, 2006).

The cost of treatment, a poor doctor-patient relationship, the need to take several pills and the adverse effects of medication are other causes of non-adherence to treatment (PIMENTA et al., 2007).

In this way, inadequate BP control may be related to lack of adherence, a simple, more important determinant of anti-hypertensive treatment and not refractoriness, and should therefore be thoroughly assessed and excluded when characterizing ARH (SOUZA, 2009).

1.2.1.3 Inadequate prescription of antihypertensive drugs

The initial drug treatment decision is based on two criteria: the level (degree) of BP and the cardiovascular risk. Each of the different classes of antihypertensive drugs has specific properties, with advantages and disadvantages (SOUZA, 2009).

Knowledge of the categories or antihypertensive drugs is important for planning monitoring programs and evaluating the impact of treatment and non-adherence, since the choice of drug treatment depends on the patient's profile, the type of AH, the presence or absence of target organ damage and associated comorbidities (LESSA, 2006).

Doctors' lack of knowledge of antihypertensive drugs, their resistance to "breaking clinical protocols" and not introducing individualized prescribing, as well as the lack of safety in modifying doses or introducing a new class of drug, creates a barrier between achieving the goal of therapy and breaking the paradigm of better preparation of the clinical staff. This aspect is important in order to direct correct and individualized therapy, minimizing the appearance of unwanted results and eliminating unnecessary pharmacotherapy, which contributes to improving adherence and consequently promoting a better quality of life for patients.

1.2.1.4 Hypertension and the white coat effect

The white coat effect is defined as a clinical condition in which systolic and diastolic blood pressure measured in the doctor's office are persistently elevated, in relation to lower blood pressure levels recorded by ambulatory blood pressure monitoring (ABPM) during the surveillance period or by home blood pressure monitoring (HBPM). It can also be defined as the presence of a difference of more than 20 mmHg in systemic blood pressure and 10 mmHg in diastolic blood pressure between the levels obtained by measuring blood pressure in the clinic and those recorded by ABPM during the surveillance period or by HBPM (GUEDIS et al, 2008).

White coat hypertension or isolated office hypertension is characterized when there is a persistence of high blood pressure levels (≥ 140/90 mmHg) in the office measurement, while the BP on ABPM or ABPM is at acceptable levels (GIRIOLI et al, 2008). An average vigilance BP <135/85 mmHg is considered normal for both the ambulatory and home monitoring methods (PIMENTA et al., 2007).

According to Pimenta et al. (2007), the prevalence of the white coat effect ranges from 20% to 40% among the general population of hypertensive patients and may be even higher in ARH, with the white coat effect occurring more commonly in women and the elderly.

1.2.1.5   Contributing Factors

1.2.1.5.1   Drug interactions

Of all the hypertensive agents, non-steroidal anti-inflammatory drugs (NSAIDs), whose use goes beyond the realm of medical prescriptions, are consumed indiscriminately and sold in pharmacies without any requirement or care. Even the specific ones, i.e. cyclooxygenase-2 (COX-2) inhibitors, inhibit renal vasodilator prostaglandins and increase the vasoconstrictor response to endogenous vasopressor substances, causing sodium and water retention and expansion of extracellular volume, with a consequent rise in blood pressure, and these effects occur more frequently in the elderly and in patients using ACE inhibitors and diuretics (RODRIGUES *et al.*, 2004).

Tricyclic antidepressants antagonize the hypotensive effects of adrenergic blocking drugs (e.g. guanethidine) by preventing the uptake of these antihypertensive drugs into adrenergic nerve endings. Similar interactions have also been observed with clonidine and methyldopa (CIARONI, 2005).

Since the beginning of the 20th century, epidemiological studies have shown an association between hypertension and alcohol consumption, which is considered a common cause of reversible BP elevation. The contribution of alcohol to the prevalence of hypertension in the population varies according to the amount ingested and the population being studied. The consumption of doses greater than 30 ml of ethanol per day is related to an increase in BP (CUSHMAN, 2001).

The effects of alcohol on BP are apparently not mediated by structural alterations, but by reversible functional vascular changes, involving the sympathetic nervous system, vasoactive substances and alterations in the cellular transport of electrolytes. Resistance to anti-hypertensive therapy has also been linked to alcohol consumption, through direct interference with the effects of hypotensive drugs or associated with pseudo-resistance, caused by the poor adherence to treatment observed in alcoholics (THADHANI et al., 2002).

A relatively new finding in epidemiological research is the observation that, at least in some countries, people who frequently drink six to eight doses of alcohol in one day (even if not every day, usually only at weekends) have an increased risk of sudden death due to cardiovascular causes. This epidemiological finding is reinforced by animal research results which show that excessive intake does not increase HDL and increases LDL, increases the risk of thrombosis, can cause changes in the myocardium and conduction tissue which facilitate arrhythmias and lowers the threshold for ventricular fibrillation. As in Brazil, excessive abuse at weekends, holidays and parties is very common, this advice not to exceed the amounts of low-risk consumption, including explaining the facts described above, is extremely important (BRITTON and MCKEE, 2000).

The act of smoking a cigarette causes a transient rise in BP, and the duration of the effect on the rise in blood pressure levels is influenced by the number of cigarettes consumed per day. Experimental studies in humans and animals have shown that the mechanism by which smoking promotes these changes is associated with the effect of nicotine on the release of neuronal and adrenal catecholamines, which increase heart rate, systemic volume, myocardial contractility, promoting systemic vasoconstriction and increased flow to skeletal muscles. Nicotine acts on the hypothalamic-pituitary axis, stimulating the secretion of corticotrophin-releasing factor, increasing the levels of endorphin, adrenocorticotrophic hormone (ACTH), vasopressin and corticosteroids in proportion to the plasma concentration reached (NATIONAL INSTITUTE OF HEALTH, 2006).

1.2.1.5.2   Volume overload

The main system that regulates BP over the long term is the Renin Angiotensin Aldosterone System (RAAS). Together with the Atrial Natriuretic Peptide (ANP) and the atrial and renal pressure receptors, they control the balance between oral fluid intake, renal excretion and extrarenal loss, thus determining the body's regular blood volume.

The volume overload that accompanies a positive body sodium balance is an important cause of ARH. Several common situations are associated with this condition, such as excessive salt intake leading to a rise in BP in hypertensive patients receiving antihypertensive medication. When volume overload is likely, assessing sodium excretion in the 24-hour urine allows us to rule out a false salt restriction reported by the patient (TOLEDO, 2006).

1.2.1.5.3   Obesity and insulin resistance

Obesity is associated with stimulation of the sympathetic nervous system, salt retention and obstructive sleep apnea. It is therefore associated with higher blood pressure levels, resistance to antihypertensive treatment and a reduction in weight significantly reduces BP (VI BRAZILIAN guidelines ON ARTERIAL HYPERTENSION, 2010).

Patients with a Body Mass Index (BMI) $\geq$30 kg/m^2 are 50% more likely to have uncontrolled BP than those with a normal BMI (<25 kg/m2). A cross-sectional study of 45,125 patients revealed that, when compared to those with a normal BMI, patients with a BMI >40 kg/m2 had a three times greater risk of requiring the use of three antihypertensive drugs and a five times greater risk of requiring four drugs for adequate BP control (BRAMLAGE et al., 2004). Therefore, weight loss should always be sought in individuals with ARH who are overweight or obese.

The available data points to a clear beneficial effect of regular physical activity in reducing BP (CALHOUN et al., 2008). Aerobic exercise has a direct effect on reducing blood pressure and improves the metabolic profile. Resistance exercises also seem to have a beneficial effect on blood pressure and should complement aerobic activities (VI BRAZILIAN GUIDELINES ON ARTERIAL HYPERTENSION, 2010). Thus, resistant hypertensive patients should be encouraged to perform light to moderate physical activity after medical assessment. A training session should not be started if systemic and diastolic blood pressures are above 160 and/or 105 mmHg, respectively.

Another important factor is the activity of plasma renin (PRA), which is increased in obese people, independently of sodium retention and increased extracellular volume. The role of angiotensin II (AngII) is reinforced by the efficacy observed in the treatment of young obese hypertensive patients with ACE inhibitors (ROCCHINI, 2002).

Obese patients with ARH also exhibit a higher degree of insulin resistance, suggesting that insulin-induced hypertrophy of the smooth muscle of resistance vessels is responsible for the increase in peripheral vascular resistance in obese patients with insulin resistance and AH. Finally, obesity and hyperinsulinemia block the effectiveness of anti-hypertensive drugs, thus contributing to resistance to drug treatment for AH (GIRIOLI, 2009).

1.2.1.5.4   Secondary hypertension

The possibility of a secondary cause is one of the fundamental points in the assessment of hypertension. In ARH, however, greater emphasis should be placed on the search for secondary causes, as well as on specific prophylaxis in suspected cases. Table 3 shows the main causes of secondary hypertension, suggestive symptoms and the main diagnostic methods for investigating these conditions (ARQUIVO BRASILEIRO DE CARDIOLOGIA, 2012).

Table 3. Main causes of secondary hypertension, symptoms and signs, initial screening tests and diagnostic confirmation.

Cause	Suggestive Clinical Findings	Screening tests	Advanced pedantry
SAOS	Snoring, sleep apnea episodes, daytime sleepiness. Obesity. Short neck	Bertim's questionnaire Epwcrtr sleepiness scale	Pofcssonography (rxfrce of apne*a- hipopne*a > 5 eventcs.ho*· a)
Primary Aklosteronism	Spontaneous or drug-induced hypocayphnia, paresthesia	Aldosterone ratio[1] renin > 30 (Rerwia < 1 and Aldo >12)	Computed Tomography (nodule or peplasia). Fudrocrystisxxa test. "Saline fusion" test, aldosterone dosage in renal veins by catheterization
Renal disease Стопка	Facial edema, uremic ha tc, aneπ⅛a, presence of diabetes or family history of nephropathy	Creati ni nenia, glomerular filtration rate estimated by formulas (<60 mb'min). rrкroalbuminuria. prote⅛iuria.	Renal ultrasound (signs of parenchymatous nephropathyJ
Renovascular hypertension	Abdominal murmur, Creatinine elevation > 30% with ACEI use. ARB or RDI. hypertension in the young or elderly	Doppter of renal arteries (peak velocity > 150 crr⅛· s renal velocity ratio aorta ≥ 3), ançpotomographyI' ançporessonància of renal arteries	'Renal arteriography (60% air damage / 20 mmHg Iranstesionab gradient)
Custwig Syndrome	ua fades, gba striae woiâceas, central obesity, hirsutism	24-hour urinary cortisol. Plasma cortisol expression test after low-dose dexamethasone (Ovemiigfttf), cort salivary sol "otumo	Computed tomography of the adrenal glands and,' or magnetic resonance imaging of the pituitary gland
Pheochromatoma	Ketalea, palpitations, sweating, tachycardia hypotension Orthostatic syncope	Plasma Metanephrines, Irenic Mephanephrines, Plasma Catecholamines	Magnetic resonance imaging of adrenals MIBG scintigraphy, PET scan
Coarctation of the aorta	Reduction of pulses in hands Pressure difference greater than 20 mmHg between arms and feet, body on back	Magnetic resonance angiography of the aorta, echocardiogram	Acrtoyafia
Hyperthyroidism **and** DCthyroidism	Tachvbradicarcka or, increased sensitivity to heat/cold, myxedema, arrhythmia or coughing, menstrual alterationsIE	TSH, free T4	Reode ultrasound
Substances that can cause arterial blood pressure to rise		Investigate the use of non-narcotic analgesics. non-steroidal anti-inflammatory drugs, Corticosteroids, Smpatcomimetic agents {decongestants, anorexigens, cocaine), "stimulants (methylphen·date, dexmethylphen*date. dextroamphetamine, amphetamine. methamphetamine, moda find), alcohol, oral Ccntraceptives. Ciciosponna. entropoein, licorice etvas (ephedra mahuang)	

SAGS: obstetric sleep apnea syndrome. ACEI: coronary enzyme inhibitor. ARB: angiotensin receptor blocker; IDR: direct renin inhibitor; TSH: thyroid stimulating hormone.

1.3 Matrix Metalloproteinase (MMP) and Tissue Inhibitor of Metalloproteinase (TIMP)

1.3.1 Concepts, types and definitions

The degradation of the extracellular matrix (ECM) is an important mechanism for tissue development, morphogenesis, repair and remodeling. Under normal physiological conditions, this process is precisely regulated, but when it becomes unbalanced, it promotes the emergence of some diseases, including cardiovascular diseases, which are the result of uncontrolled remodeling of the extracellular matrix of the myocardium and vasculature (NAGASE; VISSE; MURPHY, 2006).

Since the first description of matrix metalloproteinases (MMPs) in 1962, the family of these enzymes has grown to include around 28 members and has become a potential drug target for various therapeutic indications, i.e. studies have focused on MMPs as biomarkers, mainly for cardiovascular diseases (KUPAI et al., 2010).

According to Phillippe and colleagues (2002), MMPs are a family (approximately 28 members) of zinc-dependent enzymes (endoproteases) capable of degrading components of the extracellular matrix including the basement membrane, collagen, elastin and others. They are associated with various physiological processes such as morphogenesis, angiogenesis and tissue repair. They are divided into 6 groups according to table 4.

Table 4. MMP category and classification.

Enzymes (Groups)	Classification	MMP S[1]
Collagsnase	Collagenase 1	MMP 1
	Collagenase 2	MMP 8
	Collagenase 3	MMP 13
	Collagenase 4	MMP 18
Gelatinase	Gelatinase A	MMP 2
	Gelatinase S	MMP 9
Stromelysin	Stromelysin 1	MMP 3
	Stromelysin 2	MMP 10
	Stromelysin 3	MMP 11
Matrilysin	Matrilysin 1	MMP 7
	Matrilysin 2	MMP 28
Membrane-type MMPs	Transmenbrana	MMP 14
		MMP 15
		MMP 18
		MMP 24
		MMP 17
	GPI Àncora	MMP 25
Other	Macrophage Elastase	MMP 12
	Enamelysin	MMP 19
	No name	MMP 20
	XMMP *(Xenopus)*	MMP 21
	CA-MMP	MMP 23
	CMMP *(Gallus)*	MMP 27
	Epilysin	MMP 28

According to Visse and Nagase (2003), 24 different MMPs have been identified in vertebrates, 23 of which are found in humans.

Many MMPs are secreted by a variety of cells and can contribute to circulating MMP levels. These cells include fibroblasts, macrophages, endothelial and other cells, eosinophils, neutrophils, smooth muscle cells and cardiomyocytes (CHASE and NEWBY, 2003) (figure 1).

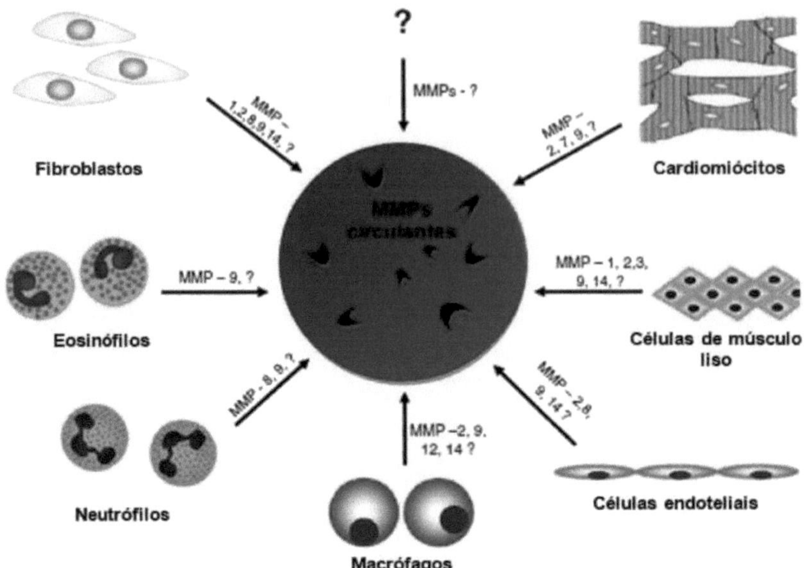

Figura 1.   Potential cellular and tissue sources of circulating MMPs found in plasma (Adapted from FONTANA et al., 2012).

Typical MMPs consist of a pro-peptide of around 80 amino acids, a metalloproteinase catalytic domain of around 170 amino acids, a peptide ligand of varying length (also called the Fhinge region) and a hemopexin (Hpx) domain of around 200 amino acids (NAGASE; VISSE; MURPHY, 2006).

The MMP family is characterized by the presence of conserved protein domains: a pro-domain, an active domain and a binding domain containing Zn^2+. All MMPs except MMP-7 and MMP-26 contain an additional carboxyterminal Hpx domain. Membrane-type MMPs (MT-MMPs) are bound to membranes via a hydrophobic carboxy-terminal anchor.

In the case of MT-MMP types 1, 2, 3 and 5, this hydrophobic anchor is a transmembrane domain and these MT-MMPs also contain a short intracellular domain, while MT-MMP- 4 and 6 are anchored to the membrane by a glycosyl-phosphatidyl-inositol. MMP-2 and MMP-9 have a fibronectin gelatin-binding domain composed of three fibronectin repeats inserted between the active domain and the site of the Zn-binding $domain^2+$, while MMP-9 contains an additional Ser/Tre/Pro domain rich in type V collagen. Compared to the other MMPs, MMP-9 is structurally one of the most complex members of the family. The Zn-binding $domain^2+$ of human MMP-9 contains a conserved sequence AHEXGHXXGXXH, in which the three histidines are responsible for coordinating the catalytic ion Zn^2+. Together with the active domain, the Zn^{2+} binding domain forms the enzyme's active center, which is essential for its enzymatic activity (PHILLIPE et al., 2002) (figure 2).

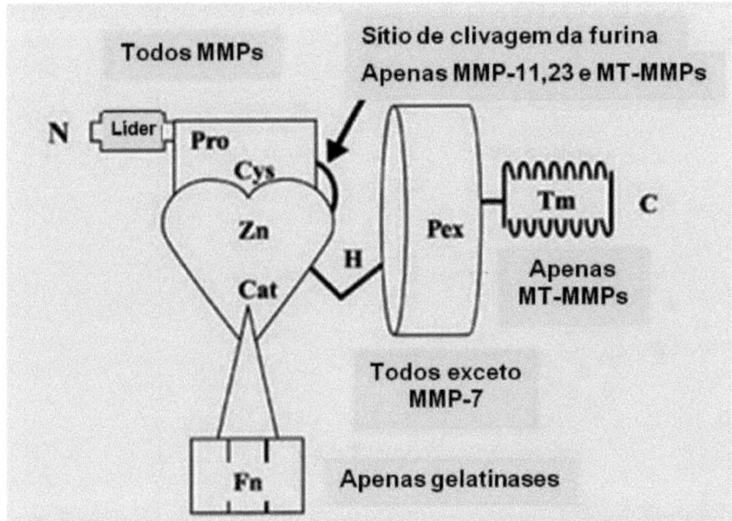

Figura 2.   Structure of MMPs. The structure of the MMP domain is shown, including the N-terminal end (N), the leader sequence (lead), pro domain (Pro), catalytic domain (Cat), fibronectin inserts (Fn), hinge region (H), hemopexin domain (Pex), transmenbrane domain (Tm) and C-terminus (C).

"In vivo" the activity of MMP's is under strict control at various levels, through transcription and post-translation mechanisms, regulated by cytokines. MMPs can be activated by reactive oxygen species (ROS) or proteases. Activated MMPs cleave ECM components rather than matrix substrates (e.g. big endothelin-1/big ET-1 and beta2-adrenergic receptors). The interactions of MMPs with their endogenous tissue inhibitors of metalloproteinases (TIMPs) in the ECM and alpha-2-macroglobulin in the serum, called natural inhibitor proteins, contribute to the regulation of MMP activity. These enzymes are generally expressed in very low quantities, in their latent form, called pro-MMP (FONTANA et al., 2012; KUPAI et al., 2010) (figure 3).

TIMPs are homologous proteins referred to as specific inhibitors of MMPs that participate in controlling the activity of these enzymes in tissues. Four types of TIMP have been identified in vertebrates (TIMP-1, TIMP-2, TIMP-3, TIMP-4) and their expression is regulated during development and tissue remodeling (VISSE; NAGASE, 2003; BREW, DINAKARPADIAN and NAGASE, 2000). TIMPs have two relatively small size domains, however, they have been found to exhibit various biochemical and physiological/biological functions, including inhibition of active MMPs, activation of pro-MMPs, promotion of cell growth, matrix binding, inhibition of angiogenesis and induction of apoptosis (BREW, DINAKARPADIAN and NAGASE, 2000).

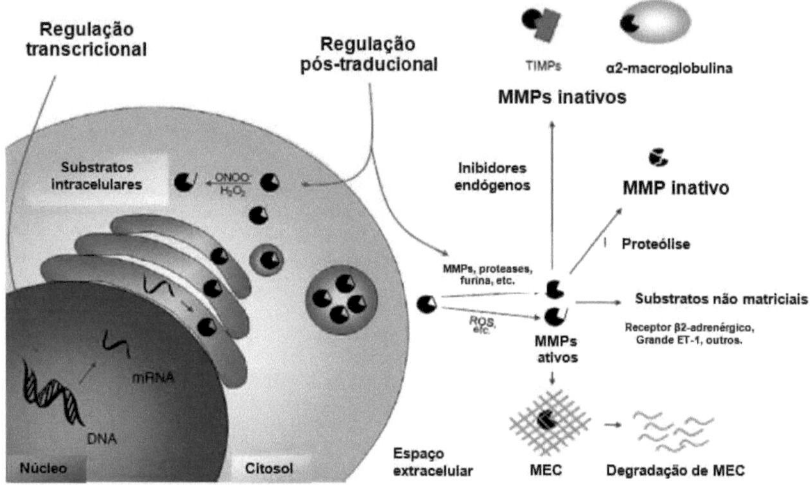

Figure 3: Regulation, expression and activity of MMPs. MMPs are regulated by transcription and post-translation mechanisms and can be activated by proteases or reactive oxygen species (ROS). Active MMPs degrade components of the extracellular matrix (ECM), leaving matrix substrates such as beta-adrenergic receptors and large endothelin-1 intact. The inhibition of MMP's occurs through the interaction of this enzyme with its natural endogenous inhibitors, such as the TIMP's present in the tissue and the alpha-2 macroglobulin present in the plasma (Adapted from FONTANA et al., 2012).

1.3.2 Actions on the cardiovascular system and cardiovascular diseases

According to Favarelli (2001), maladjustments in the delicate balance between mechanisms that maintain the physiological conditions of the vasculature can be summarized as follows: 1) hemodynamic disturbances, represented mainly by shear stress; 2) regulation of vascular tone by vasoactive factors: vasodilators and vasoconstrictors; 3) proliferative and anti-proliferative activity: vascular remodeling; 4) altered platelet activity; 5) oxidative stress; 6) pro-inflammatory and anti-inflammatory activity; and 7) hemostatic and thrombolytic factors.

The rigidity of the vascular wall is determined by the relative concentration of collagen and elastin, which are regulated by the balance between the production and degradation of ECM

(ZHOU et al, 2007).

The main function of MMPs was considered to be the degradation and removal of extracellular matrix molecules from the tissue. However, it has been increasingly recognized that the breakdown of ECM molecules or cell surface molecules alters cell-matrix and cell-cell interactions and the release of growth factors that are bound to the ECM becomes available to cell receptors (NAGASE; VISSE; MURPHY, 2006).

MMPs play an important role in tissue remodeling (mainly vascular remodeling) and contribute to the development and progression of some pathological conditions, such as rheumatoid arthritis, coronary artery disease (CAD) and cancer (TANYA et al, 2004).

The proteolytic system made up of MMPs and TIMPs is involved in regulating the metabolism of ECM, contributing to vulnerability to vascular damage caused by the formation of atherosclerotic plaques (SOLINI et al., 2009).

1.4 MMP-9 and TIMP-1

1.4.1 Definition, activity, regulation and expression

MMP-9 belongs to the class of metalloproteinases called gelatinases. It was the first to be purified in human macrophages and its expression is controlled by growth factors, chemokines and other stimulatory signals. It is secreted in the form of an inactive precursor (pro-MMP-9), i.e. it is secreted as a pro-enzyme and needs to be activated in order to exert its catalytic activity (PHILIPPE et al, 2002).

MMP-9 (gelatinase B, 92-kD collagenase type IV) is an important member of the metalloproteinase family and is capable of degrading gelatin, collagenase-degraded fragments of collagen and collagen type IV, which form part of the basement membrane (PHILIPPE et al, 2002).

According to Phillipe et al. (2002), MMP-9 is synthesized and secreted as a zymogen or proenzyme, which remains inactive if not activated by removal of the propeptide. This propeptide contains the conserved sequence PRCGXPD, of which Cys is coordinated with the catalytic ion Zn^2+. After disruption of coordination, for example by proteolysis of the propeptide, a conformational change occurs and the Zn^2+ domain becomes accessible by a water molecule and hydrolyzes to the substrate, resulting in activation of the enzyme.

MMP-9 activity is regulated through five mechanisms: gene transcription, activation, secretion, inhibition and glycosylation. In most cells, transcription of the MMP-9 gene is inducible, and after translation the enzyme is immediately secreted via the normal secretory pathway. It is normally secreted together with varying amounts of its inhibitor TIMP-1 and MMP-2 (PHILLIPE et al., 2002).

Once MMP-9 is secreted and activated, its activity is regulated by degradation or inhibition, which is inhibited by α2-macroglobulin, the universal protease inhibitor present in human serum. However, TIMPs are more specific and more important in regulating MMP activity. TIMPs are stable glycoproteins with a relative molecular weight of 20 to 30 kDa and contain six conserved disulfide bridges. These disulfide bridges define six protein loops, so that the first three consist of an amino-terminal domain and the others comprise a carboxy-terminal domain. These domains fold independently on top of each other, and the amino-terminal domain can carry out its function independently, i.e. without the carboxy-terminal domain. Four different TIMP genes and proteins have been described in humans, of which TIMP-1 binds with high affinity to MMP-9, this inhibitor being glycosylated at two sites. TIMP inhibition follows slow binding kinetics and is highly complex, while different TIMP binding sites are present in both gelatinases. Not only can activated MMP-9 bind to different TIMPs, but the proenzyme is also able to bind to TIMP-1. The interaction between pro-MMP-9 and TIMP-1 seems to occur mainly through the C-terminal domains of both molecules, due to a C-terminal deletion mutant of TIMP-1 not binding to pro-MMP-9, just as the C-terminal mutant of MMP-9 does not bind to TIMP-1. Complexes of pro-MMP-9 and TIMP-1 are able to inhibit other MMPs by forming an MMP-9/TIMP-1/MMP complex, indicating that inhibition by the N-terminal domain of TIMP-1 is available for interaction with the pro-MMP-9/TIMP-1 complex. Inhibition of activated MMP-9, on the other hand, occurs through interaction between the N-terminal domains of TIMP-1 and the active site of the enzyme, as the C-terminus of TIMP-1 deletion mutants retains its inhibitory activity against MMP-9. The C-terminus of MMP-9 deletion mutants lacking the type V collagen Hpx domain are less effectively inhibited by TIMP-1, indicating the involvement of part of the C-terminus in this process. The C-terminal domains appear to be responsible for the high-affinity interaction with TIMP-1 with a dissociation constant at the nanomolar level. In contrast, the N-terminal domains are responsible for the low-affinity interaction with a dissociation constant at the micromolar level (PHILLIPE et al., 2002).

1.4.2 Participation in cardiovascular diseases

All the events that disrupt the normal hemodynamics of the cardiovascular system related to hypertensive disease are linked to the process of stiffening of large arteries caused by the loss of elasticity of the arteries due to the action of enzymes that degrade the extracellular matrix in these structures, including MMP-9, as discussed above (MARTINS et al., 2011).

This process of physiological alteration has several adverse consequences, including an increase in pulse pressure and changes in shear stress, which promote vascular and cardiac remodeling, thus increasing cardiovascular risk (YASMIN et al.; 2005).

MMPs, especially MMP-9, are widely expressed in vascular and cardiac tissue and are involved in cardiac rupture and dysfunction after myocardial infarction (INTEGRAN AND SCHIFFRIN, 2001).

Clinical studies have shown an increase in circulating MMP-9 associated with left ventricular remodeling after acute myocardial infarction (KELLY et al., 2007).

MMP inhibition may provide an effective therapeutic strategy in the acute phase of acute myocardial infarction, possibly preventing heart damage (GUIMARAES et al., 2010).

Alterations in TIMP concentration have been reported in cardiovascular diseases. TIMP-1 was associated with left ventricular remodeling in patients without a history of heart failure or myocardial infarction and after acute myocardial infarction (KELLY et al., 2008). A positive association has been reported between increased TIMP-1 and adverse cardiac events and mortality (VELAGALETI et al., 2010). Increased cardiac TIMP-1 expression was related to interstitial fibrosis and cardiac dysfunction (HEYMANS et al., 2005).

High expression of MMP-9 has been associated with destabilization of the coronary plaque and with remodeling of the outer artery and aneurysm formation. This phenomenon presumably occurs as a result of excess degradation of extracellular matrix components. MMP-9 substrates include denatured collagen (gelatin), type II, IV and V collagen, elastin, entactin, and vitronectin. Excessive degradation of important components of the extracellular matrix can influence aortic stiffening (ZHOU et al, 2007).

Increased MMP-9 levels have been linked to acute cardiovascular events and/or AH in individuals without cardiovascular disease. In hypertensive patients, increased MMP-9 activity can result in the degradation of elastin, raising it in relation to collagen and non-elasticity, while reduced TIMP-1 activity can lead to an accumulation of weakly cross-linked, immature and unstable fibrillar degradation products, which results in collagen shedding (ONAL et al.; 2009).

Aortic stiffness is directly proportional to circulating levels of MMP-9 and elastase, not only in isolated systemic hypertension (ISH), but also in younger, apparently healthy individuals. This suggests that elastases, including MMP-9, may be involved in the process of arterial stiffening and the development of SIH (YASMIN et al.; 2005).

Growing evidence suggests that MMPs can promote an increase in BP through direct interaction with receptors or modulation of vasoactive regulatory peptides (FERNANDEZ-PATRON et al., 1999).

Furthermore, it has recently been shown that MMPs can cleave the extracellular domain of e2-adrenergic receptors in hypertensive rats and therefore suppress β-agonist-induced vasodilation. Although these studies are opening up a whole new field of investigation, it is clear that MMPs can degrade many other substrates, including intracellular targets in the cardiovascular system (CHOW et al., 2010).

These experimental studies have stimulated further research into MMPs and their tissue inhibitors as potential biomarkers in hypertension. It is possible that the circulating concentration of these enzymes and their inhibitors may be associated with cardiovascular disease, hypertension complications and prognosis, and may therefore be useful in clinical practice (AHMED et al., 2006).

1.5 Justification

It is known that MMP-9 and TIMP-1 are involved in some cardiovascular processes such as vascular restructuring, which in turn are directly associated with ARH and may be playing an important role in its cause and influencing BP control in these patients.

By comparing the levels of MMP-9 and TIMP-1 in patients with HARNC and HARC, we can identify the distinct characteristics of these subgroups that may contribute to resistance to antihypertensive therapy. It is important to note that there are no studies in the literature reporting results similar to those proposed in this study.

1.6 Hypotheses

It is expected that plasma levels of MMP-9 will be higher in these patients, especially in the HTRNC group.

A positive correlation is expected between MMP-9 and TIMP-1 levels and OPV and PP in these patients.

2 OBJECTIVES

2.1 General objective

To compare plasma levels of MMP-9, TIMP-1 and the MMP-9/TIMP-1 ratio between controlled resistant hypertensives (CRHT) and uncontrolled resistant hypertensives (NCRHT).

2.2 Specific objectives

To check the correlation between plasma levels of MMP-9, TIMP-1 and the MMP-9/TIMP-1 ratio and clinical variables (gender, age, BMI, SBP, DBP, PP, OPV, total cholesterol, LDL, HDL, triglycerides, urea, creatinine, blood glucose, uric acid, aldosterone and renin) in all resistant hypertensives (RHT), controlled resistant hypertensives (CRHT) and uncontrolled resistant hypertensives (NCRHT).

3 MATERIALS AND METHODS

3.1 Patient population

A total of 76 individuals diagnosed with ARH who were being followed up at the Resistant Hypertension Outpatient Clinic at the University of Campinas - UNICAMP/SP took part in this study. Of all the patients, 35 were classified as HTRNC and 41 were classified as HTRC.

All the patients had a family history of the case (ARH) and underwent a physical examination, electrocardiogram and laboratory tests to assess biochemical parameters. Patients with secondary forms of hypertension, kidney failure, ischemic heart disease, liver disease, peripheral vascular disease, stroke, smoking or any other serious illnesses were excluded from the study. Patients were investigated for adherence to the stipulated clinical treatment and optimization of antihypertensive therapy.

Ambulatory BP monitoring (Spacelabs 90207, Spacelabs Inc, Redmon, WA, USA) was used only to exclude cases of pseudo-resistance and as an auxiliary method for evaluating the effectiveness of antihypertensive treatment and for characterizing HARC and HARNC. This study was approved by the Ethics Committee of the State University of Campinas, São Paulo, Brazil, under protocol number 222 of 2011. All participants were aware of the nature of the research study and signed a written informed consent form before enrolling in the research.

3.2 Study design

All patients with ARH were assessed. Guidance was given on controlling salt in the diet, which was confirmed by measuring urinary sodium excretion. Patients underwent casual blood pressure measurement, pulse wave velocity (PWV), plasma MMP-9 and TIMP-1 dosage, determination of Plasma Aldosterone Concentration (APC) and Plasma Renin Activity (PRA).

3.3 Laboratory evaluation

Blood samples for measuring ARP, CPA and other biochemical parameters were taken at 08:00 hours after an overnight fast. During this time, the volunteers rested in the supine position for 8 hours (including time for Ambulatory Blood Pressure Monitoring - ABPM), followed by 1 hour in the upright position in an air-conditioned room (22-24 °C). PRA was measured by a commercial laboratory (Mayo Clinic Laboratories, Rochester, Minnesota, USA), using standard techniques. PRA levels were measured by radio immunoassay.

3.4   PA office measurements

With the patients in a sitting position, the BP levels after resting for 5 minutes and the office BP level were obtained according to the American Heart Association. BP was measured twice using an automatic digital BP monitor (OMRON Healthcare Inc., Bannockburn, IL, USA). Final BP was determined by averaging the two measurements. Pulse pressure was calculated as the difference between systolic BP and diastolic BP (SBP - DBP).

3.5   Measurement of aortic PWV

Aortic PWV was measured using the foot-to-foot velocity method with patients in the supine position, with prior validation by Complior SP equipment and software (Artech-Medical, Paris, France). Waveforms were obtained transcutaneously using the right common carotid and femoral arteries simultaneously for a minimum period of 10 to 15 seconds. The time delay (t) was measured between the two waveforms, and the distance (D) traveled by the waves was measured directly between the femoral recording site and the suprasternal notch, minus the distance of the suprasternal notch from the carotid recording site. OPV was calculated as D(m) / t(s).

Three consecutive readings were taken, and the final OPV was stipulated by the average of the readings.

3.6   Piasmatic dosage of MMP-9 and TIMP-1

Quantification of piasmatic MMP-9 in the patients' blood samples was carried out using the DuoSet® ELISA human MMP-9 method (RnD System), following the standard protocol specific to the technique. The quantification of piasmatic TIMP-1 was obtained using the DuoSet® ELISA human TIMP-1 (RnD System) method, following the specific protocol for the technique.

3.7   Statistical analysis

The Statistical Analysis System, version 3.02 (Prism GraphPad Inc, 2000), was used for all statistical analyses in the study. All values are expressed as mean ± standard deviation. Normality of distribution was assessed, with clinical variables with asymmetric distribution boing analyzed using the non-parametric student's t-test and clinical variables with normal distribution being analyzed using the parametric student's t-test. Significant differences between study subgroups were determined using t-test analysis, and correlations were expressed by Pearson's correlation coefficient (normal distribution) or Spearman's correlation coefficient (non-normal distribution). P-values ≤ 0.05 were considered statistically different.

4 RESULTS

4.1   Patients' general characteristics (t-test).

The genetic characteristics of the study groups are listed in Table 5.

With regard to gender, there was no statistical significance between the groups. The majority of patients were female, accounting for 22 of the 35 HTRNC patients and 27 of the 41 HTRC patients (63 and 66%, respectively).

Statistically significant differences were found in the parameters SBP, DBP, PP, OPV and HDL. Significantly, SBP, DBP and PP levels were 14%, 12% and 16% higher respectively in the HTRNC compared to the HTRC group. HDL levels were 15% higher in HTRC. The other parameters were not different between the groups.

Table 5. General characteristics of patients

	HTRNC (N= 35)	HTRC (N= 41)
Female (%)	S3 %	66%
Age (years)	57 ± 11,3	61 ±9,7
BMI (Kg/m)^2	30,0 ±4.4	30.1 ±4.4
SBP (mmHg)	157,9 ± 19,7*	136,2 ± 14,5
DBP (mmHg)	90,7 ± 14,3 '	79.9 ± 7,4
PP (mmHg)	67.2 ± 13,4 *	.56,3 ± 13,3
OPV (mmHg)	11,9 ± 1,8 *	10,6 ± 1,3
Total Cholesterol (mg/dl)	203,6 ±50.4	202,1 ±38,9
LDL (mg/dl)	126.1 ±38.3	125,4 ±35,0
HDL (mg/dl)	41,2± 12.0 *	48,0 ± 14.0
Triglycerides (mg/dl)	180,9 ±96,2	149,2 ± 66,5
Urea (mg/dl)	38,1 ± 11,8	35,9 ± 7,5
Creatinine (mg/dl)	1.0 ± 0,2	0,9 ±0,2
Blood glucose (mg/dl)	125.4 ±54,1	106,7 ± 34,2
Uric acid (mg/dl)	5,9 ± 1,6	5,8 ± 1,5
Aldosterone (pg/mL)	109,7 ± 82,0	101.1 ± 70.5
Renin (pq/mL)	22,4 ± 19,6	21,2 ± 18,2

Abbreviations: BMI, body mass index; SBP, systemic arterial pressure; DBP, diastolic arterial pressure: PP. pulse pressure; PWV. Pulse wave velocity; LDL, low density lipoprotein; HDL, high density lipoprotein; HTRNC, uncontrolled resistant hypertension; HTRC, controlled resistant hypertension.

Values expressed as mean = SD.

* P<0.05 vs HTRC.

Table 6 shows the percentage distribution of the classes of antihypertensive drugs used by the subgroups of patients with ARH, given that these patients use three or more classes of these drugs.

Table 6. Class of antihypertensive drugs used by patient subgroups

	HTRNC (N= 35)	HTRC (N= 41)
Spironolactone (%)	32%	23%
Diuretics from other classes (%)	100%	100%
Beta-blockers (%)	52%	46%
ACEI (%)	27%	33%
BRA II (%)	77.3% *	35%
BCC (%)	65.9% *	29%
Antı-hıpertensıvo de Acción Central (%)	11%	4%

Abbreviations ACEI. angiotensin-converting enzyme inhibitors; BFiA II, angiotensin II receptor blockers BCC Calcium channel blockers

* P<0.0001 vs HTRC.

In relation to angiotensin II receptor blockers (ARB II) and calcium channel blockers (CCB), there was a significant difference (P=<0.0001), and the use of ARB II and CCB was higher in HTRNC (77% and 66%, respectively) than in HTRC (35% and 29%, respectively). No significant differences were found in the other classes of antihypertensive drugs.

There were no statistically significant differences in the levels of MMP-9, and the ratio (MMP-9/TIMP-1), between the groups (figure 4), however, a significant difference was observed in relation to TIMP-1 between the groups (P=0.0017). Although MMP-9 levels between the groups (HTRNC and HTRC) were similar (235.8 ± 136.0 ng/mL vs 230.1 ± 122.6 ng/mL respectively, figure 4A) TIMP-1 values were higher (19%) in the HTRNC compared to the HTRC group (504.1 ± 411.2 ng/mL vs 407.3 ± 249.1 ng/mL respectively, figure 4B). As for the ratio, similar values were observed in HTRC patients (0.80 ± 0.60) compared to HTRNC patients (0.70 ± 0.50), i.e. a 12.5% reduction in HTRNC (figure 4C).

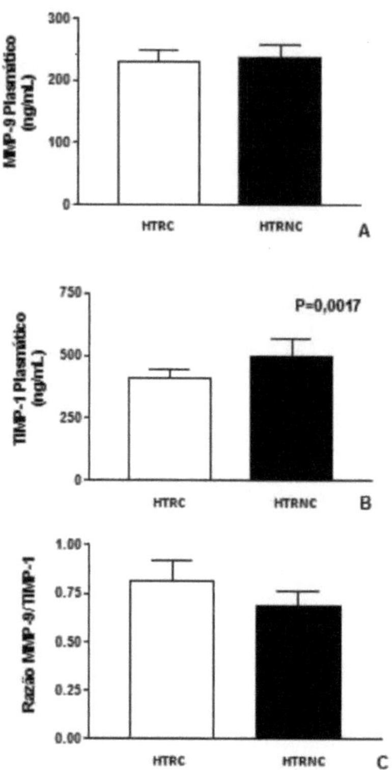

Figure 4: Circulating values of MMP-9, TIMP-1 and the MMP-9/TIMP-1 ratio between the HTRC - controlled resistant hypertensive and HTRNC - uncontrolled resistant hypertensive groups.

4.2 Analysis of correlations between MMP-9, TIMP-1 and the MMP-9/TIMP-1 ratio with clinical variables.

Tables 7 to 11, which will be presented throughout the text, represent the correlations between the plasma levels of MMP-9, TIMP-1 and their ratio with the blood pressure and biochemical parameters of the study group, with their respective significance values (P value) and correlation (Pearson or Spearman).

In relation to age, significant differences were found when comparing TIMP-1 with the HTRC subgroup (P=0.05), which also occurred with the MMP-9/TIMP-1 ratio and the HTRNC subgroup (P=0.02), both showing a positive correlation (r=0.30 and r=0.37, respectively) and all HTR (P=0.04), with the latter showing a negative correlation (r=-0.23). With regard to MMP-9, no significant differences were found (Table 7).

28

Table 7. Correlation between age and biomarkers

CLINICAL CHARACTERISTICS	HTRC	HTRNC	ALL	BIOMARKERS
	r=0.12 P=0.42	r=-0,25 P=0,13	rs=0.02 P=0.84	MMP-9
AGE	r=0.30 P=0.05	r=0.10 P=0.54	r=0.14 P=0.22	TIMP-1
	rs=0.22 P=0.16	r=0.37 P=0.02	rs=-0.23 P=0.04	RATIO MMP-9/TIMP-1

Abbreviations: HTRC, controlled resistant hypertension; HTRNC, uncontrolled resistant hypertension.

When compared with MMP-9, the body mass index showed statistical significance in the HTRC subgroup (P=0.02), with a negative correlation (r=-0.36). In the MMP-9/TIMP-1 ratio, the body mass index showed a significant difference in all HTRs (P=0.03), with a negative correlation (r=-0.25). In relation to TIMP-1, no significant differences were found, as shown in Table 8.

Table 8. Correlation between BMI and biomarkers

CHARACTERISTICS	HTRC	HTRNC	ALL	BIOMARKERS
	r=-0.36 P=0.02	r=-0,14 P=0,39	r=-0.22 P=0.06	MMP-9
BMI	r-0.18 P=0.24	r=0.24 P=0.15	r=0,21 P=0,07	TIMP-1
	r=-0.27 P=0.08	r=-0.23 P=0.17	r=-0.25 P=0.03	RATIO MMP-9/TIMP-1

Abbreviations: BMI, body mass index; CRH, controlled resistant hypertension;

HTRNC, uncontrolled resistant hypertension.

Analyzing the correlation between the parameters evaluated and the biomarkers (MMP-9, TIMP-1 and MMP-9/TIMP-1 ratio) in their respective subgroups (HTRC, HTRNC and all HTR), there was no significant correlation with SBP, however, in relation to DBP, there was a significant negative correlation (r=- 0.33) when comparing the plasma levels of TIMP-1 with the HTRNC subgroup (P=0.04). In the MMP-9/TIMP-1 ratio with the HTRNC subgroup, there was a significant difference (P=0.02) with a positive correlation (r=0.37). There were no significant differences in MMP-9 and its subgroups (Table 9).

Table 9. Correlation between SBP and DBP and biomarkers

LINGUISTIC CHARACTERISTICS	HTRC	HTRNC	ALL	BIOMARKERS
SBP (mmHg)	r=0.23 P=0.14	r=-0.19 P=0.26	r=0.14 P=0.22	MMP-9
DBP (mmHg)	r=0.06 P=0.67	r=-0.07 P=0.65	r=0.14 P=0.24	
SBP (mmHg)	r=0.02 P=O₁ 85	r=-0.23 P=0.17	r=-0.04 P=0.72	TIMP-1
DBP (mmHg)	r=0.12 P=0.44	r=-0.33 P=0.04	r=-0.12 P=0.28	

SBP (mmHg)	r=0.07 P=0.64	r=0.15 P=0.38	r=0.02 P=0.85	RATIO MMP·9/TIMP·1
DBP (mmHg)	r=-0.12 P=0.42	r=0.37 P=0.02	r=0.06 P=0.59	

Abbreviations: SBP, systemic arterial pressure; DBP, diastolic arterial pressure; HTRC, controlled resistant hypertension; HTRNC, uncontrolled resistant hypertension.

No significant differences were observed in the PP and VOP parameters, as shown in Table 10.

Table 10. Correlation between PP and OPV and biomarkers

CÜNICAL CHARACTERISTICS	HTRC	HTRNC	ALL	BIOMARKERS
PP (mmHg)	r=0.20 P=O.18	r=0.19 P=0.25	r=0.08 P=O,50	MMP-9
VOP C-S (m/s)	r=0.07 P=O.63	r=-0.17 P=0.31	r=-0,13 P=O,26	
PP (mmHg)	r=-0.04 P=0.80	r=0.02 P=0.86	r=0.05 P=0.66	TIMP-1
VOP C-S (m/s)	r=0.06 P=0.67	r=-0.07 P=0.68	r=-0.007 P=0.95	
PP (mmHg)	r=0.14 P=0.35	r=-0.19 P=0.25	r=-0.02 P=0.82	RATIO MMP-9/TIMP1
VOP C-S (m/s)	r=0.21 P=0.17	r=-0.01 P=0.92	r=0.05 P=0.67	

Abbreviations: PP, pulse pressure; PWV, pulse wave velocity; HTRC, controlled resistant hypertension; HTRNC, uncontrolled resistant hypertension.

The results of the biochemical tests were very similar in both groups. With regard to the lipid profile of the patients, no significant differences were found in the correlation analyses (table 11), nor in their renal profile (table 12).

Table 11. Correlation between lipid profile and biomarkers

CLINICAL CHARACTERISTICS	HTRC	HTRNC	ALL	BIOMARKERS
Total cholesterol (mg/dl)	r=-0.04 P=O.79	r=-0.14 P=O.42	r=0.05 P=O.69	MMP-9
LDL (mg/dl)	r=0,01 P=O,90	r=-0.03 P=O.84	r=0.06 P=O.62	
HDL (mg/dl)	r=0.00 P=O.99	r=-0,07 P=O,66	r=-0,10 P=O,38	
Triglycerides (mg/dl)	r=-0.14 P=O.38	r=-0,15 P=O,36	r=0.17 P=O.13	
Total Cholesterol (mg/dl)	r=0.26 P=O.09	r=-0.18 P=O,28	r=-0,02 P=O,85	TIMP-1
LDL (mg/dl)	r=0.16 P=O.30	r=-0,13 P=O,43	r=-0,01 P=O,90	
HDL (mg/dl)	r=-0.06 P=O.68	r=-0.15 P=O,36	r=-0,14 P=O,23	
Triglycerides (mg/dl)	r=0.26 P=O.09	r=-0.10 P=0.53	r=0.02 P=O.85	
	r=-0.13 P=O.39	r=0.16 P=O.35	r=-0.009 P=O,94	RATIO MMP-9/TIMP-1
Total Cholesterol (mg/dl)	r=-0.04 P=O.78	r=0.18 P=O.29	r=0.04 P=O.71	
LDL (mg/dl)	r=0.07 P=O.63	r=0.09 P=O.57	r=0.04 P=O.71	
HDL (mg/dl) Triglycerides (mg/dl)	r=-0,24 P=O,12	r=0.30 P=O.07	r=0.004 P=O.97	

Abbreviations: LDL, low density lipoprotein; HDL, high density lipoprotein; HTRC, controlled resistant

hypertensives; HTRNC, uncontrolled resistant hypertensives.

Table 12. Correlation between renal profile and biomarkers

CLINICAL CHARACTERISTICS	HTRC	HTRNC	ALL	BIOMARKERS
Irea (mg/dL)	r=-0.09 P=0.57	r=0.16 P=0.32	r=-0,09 P=0,41	MMP-9
Creatinine (mg/dL)	r=0.20 P=0.19	r=0,01 P=0.92	r=0,11 P=0,32	
Irea (mg/dL)	r=-0,20 P=0,20	r=-0.21 P=0.21	r=-0,19 P=0,10	TIMP-1
Creatinine (mg/dL)	r=-0.05 P=0.71	r=-0.32 P=0.06	r=-0,18 P=0,11	
Lorraine (mg/dL)	r=0.06 P=0.69	r=0.24 P=0.14	r=0.12 P=0.28	RATIO MMP-9/TIMP-1
Creatinine (mg/dL)	r=0.15 P=0.34	r=0.17 P=0.32	r=0.14 P=0.22	

Abbreviations: HTRC, controlled resistant hypertension; HTRNC, uncontrolled resistant hypertension.

When comparing plasma levels of TIMP-1 with the HTRC subgroup and all HTRs, a significant difference was observed in the glycemia parameter (P=0.04 and P=0.002, respectively), with a positive correlation (rs=0.32 and rs=0.34). For the same parameter, comparing the MMP-9/TIMP-1 ratio with the subgroups of HTRC and all HTR, significant differences were found (P=0.05 and P=0.02), however, a negative correlation was observed (rs=-0.30 and rs=-0.27), as shown in table 13.

In the biochemical parameter uric acid, significant differences were found when comparing plasma levels of TIMP-1 and the subgroups of HTRNC and all HTR (P=0.01 and P=0.002), with a negative correlation in both (r=-0.47 and r=-0.35). In the MMP-9/TIMP-1 ratio with the HTRNC subgroup, there were statistically significant differences (P=0.01), with a positive correlation (r=0.44), as can be seen in table 13.

Table 13. Correlation between blood glucose and uric acid and biomarkers.

CUNIC CHARACTERISTICS	HTRC		HTRNC		ALL		BIOMARKERS
Blood glucose (mg/dl)	rs=0.01	P=0,92	r=-0,25	P=0,14	rs=0.09	P=O,44	MMP-9
Uric acid (mg/dL)	r=0,10	P=O,51	r-0,22	P=0,21	r=0,09	P=0,43	
Blood glucose (mg/dl)	rs=0.32	P=O,04	r=0,14	P=0,41	rs=0.34	P=0,002	TIMP-1
Uric acid (mg/dL)	r=-0,22	P=0,16	r=-0,47	P=0,01	r=-0,35	P=0,002	
Blood glucose (mg/dl)	rs=-0.30	P=0,05	r=-0,13	P=0,42	rs=-0.27	P=0,02	RATIO MMP-9/TIMP-1
Uric acid (mg/dL)	r=0,07	P=0,64	r=0,44	P=0,01	r=0,21	P=0,08	

Abbreviations: HTRC, controlled resistant hypertension; HTRNC, uncontrolled resistant hypertension.

In relation to piasmatic aldosterone, when the MMP-9/TIMP-1 ratio was correlated with the HTRNC subgroup, there was a significant difference (P=0.005) and a positive correlation (r=0.56). In the correlation between the other subgroups, no significant differences were observed (Table 14). Piasmatic renin in the HTRNC subgroup compared with plasma levels of MMP-9 was statistically significant (P=0.04), although it showed a negative correlation (rs=-0.34). There were no statistically significant differences in the correlation of the other subgroups (table 14).

Table 14. Correlation between aldosterone and renin and biomarkers.

CLINICAL CHARACTERISTICS	HTRC	HTRNC	ALL	BIOMARKERS
	r=0.08 P=O.64	r=0.08 P=O.64	rs=0.15 P=O.22	MMP-9
Aldosterone (pg/mL)	r=-0.31 P=O.06	r=-0,13 P=O,44	rs=-0.11 P=O.37	TIMP-1
	r=0.02 P=O.87	r=0.56 P=O,005	rs=0.18 P=O,13	RATIO MMP-9/TIMP-1
	rs=0.01 P=O.95	rs=-O,34 P=O,04	rs=0.003 P=O.98	MMP-9
Renin (pg/mL)	rs=-0,09 P=O,55	rs=0.02 P=O.92	rs=-0.05 P=O.65	TIMP-1
	rs=0.12 P=O.42	rs=0.04 P=O.79	rs=0.02 P=O.83	RATIO MMP-9/TIMP-1

Abbreviations: HTRC, controlled resistant hypertension; HTRNC, uncontrolled resistant hypertension.

5 DISCUSSION

The results obtained by the study show a greater relationship between TIMP-1 and resistant hypertension, specifically in the HTRNC group, contradicting the initial hypothesis that MMP-9 levels would be higher in this population. We found it difficult to discuss this finding, as there are no similar studies in the literature showing this data, especially in the CRHT and NRHT groups. Therefore, it can be said that TIMP-1 apparently plays an important role in resistant hypertension.

One of the limitations of this study is the difficulty in assessing these biomarkers in plasma, since their activity in tissue is different from that in biological fluids.

It is known that AH is associated with significant activation of the renin-angiotensin system, producing vascular alterations possibly resulting from increased oxidative stress and excessive activation of ECM MMP's (vascular remodeling), resulting in structural changes in the cardiac and vascular ECM, with MMP's and TIMP's participating in the vascular remodeling process, influencing these changes which may be physiological or pathological (GUIMARÂES et al., 2010).

Under physiological conditions, there is a balance between the ratio of MMPs and TIMPs. However, in pathological processes such as AH, there is an imbalance in this ratio, leading to excessive degradation of ECM proteins (MURPHY AND NAGASE, 2008) and consequently to pathological vascular remodeling (SLUIJTER et al., 2006).

According to Tanya et al. (2004), arterial stiffness is an independent predictor of cardiovascular events in the hypertensive population. Circulating levels of MMP-9 are increased in hypertension in type II diabetic patients with coronary artery disease and are associated with premature coronary atherosclerosis, with plasma levels of MMP-9 being associated with vascular stiffness and cardiovascular risk predisposition.

Recently, high expression and activity of MMP-9 have been demonstrated in vascular remodeling in hypertensive patients (AHMED et al., 2006; YASMIN et al., 2005).

Given this knowledge, we expected to find high levels of MMP-9 in the patients under study, especially in the HTRNC group, which was contradictory to the results presented. MMP-9 activation is more closely related to AH in terms of excessive degradation of ECM components, thus causing pathological vascular remodeling (GUIMARAES et al., 2010) which can influence responsiveness to antihypertensive therapy.

One explanation for not finding high levels of MMP-9 in the patients in question may be the fact that these individuals are already classified as resistant hypertensive patients, i.e. they are on antihypertensive therapy, since this therapy influences the decrease in MMP-9 activity, mainly related to the use of ACE inhibitors, as shown in some studies.

MMPs are zinc-dependent endopeptidases and have this ion in their active site. Similarly, ACE is also a zinc-dependent enzyme that is inhibited by ACE inhibitors, which are widely used in current anti-hypertensive treatment. It is suggested that the blocking mechanism of ACE inhibitors involves the binding of zinc in the active site, since some patients receiving ACE inhibitors develop dysgeusia due to zinc deficiency. Given this, it is possible that MMP-9 may also be inhibited by ACEIs through the binding of zinc in the active site. The ACEI captopril can inhibit MMP-9 activity, and this inhibition is reversible in the presence of excess zinc (SORBI et al., 1993).

Inoue et al. (2010) reported that ACEIs are effectively stabilized by hydrogen bonds and specific hydrophobic interactions at the active site of MMP-9. Based on the evidence, it is suggested that treatment with ACEIs can inhibit MMP-9 activity, which differs from our findings since the use of this class of antihypertensive was not significant between the groups.

One hypothesis to be investigated with regard to the increase in plasma levels of TIMP-1 is that it generates a response which consists of modulating or limiting the degradation of collagen and thus contributes to the development of arterial stiffness. Unlike MMP-9, some studies indicate an increase in TIMP-1 after antihypertensive treatment.

Onal and colleagues (2009) demonstrated in their study that there is a reduction in plasma levels of MMP-9 in hypertensive patients when they are undergoing antihypertensive treatment, while serum levels of TIMP-1 are increased after antihypertensive treatment. This suggests that the MMP/TIMP enzyme system may play an important role in the development of hypertensive disease in target organs.

Muzahir et al. (2004) carried out a study on MMP-9 and TIMP-1 in hypertension, observing that before treatment, circulating levels of MMP-9 and TIMP-1 were significantly higher in patients with hypertension than in normotensive controls (P = 0.0041 and P = 0.0166, respectively). Plasma levels of MMP-9 decreased and TIMP-1 levels increased after antihypertensive treatment (P = 0.035 and P = 0.005, respectively).

Lindsay, Maxwell and Dunn (2004) reported in their study that TIMP-1 levels were significantly higher in the hypertensive group compared to normotensive individuals (385 ng/mL and 253 ng/mL, P= 0.0007, respectively), so TIMP-1 levels are not elevated in hypertension alone, but only in patients with diastolic dysfunction and fibrosis, suggesting that the synthesis and release of TIMP-1 is independent of blood pressure and is probably dependent on a variety of neurohormonal factors, with TIMP-1 levels > 500 ng/mL is an accurate indicator of diastolic dysfunction and therefore of target organ damage.

In a study carried out by Ahmed et al. (2006) evaluating the profiles of MMPs and TIMPs in AH, one of the findings was that alterations in the MMP and TIMP profiles favoring a decrease in ECM degradation (a decrease in MMP-2, -9 and -13 and an increase in TIMP-1) were associated with left ventricular hypertrophy and diastolic dysfunction, and an increase in TIMP-1 predicted the presence of chronic heart failure.

These findings may explain the correlation between PAD and TIMP-1 in HTRNC patients, as demonstrated in this study, although this correlation was negative.

According to Ahmed et al. (2006), in hypertensive heart disease, if the pyosmotic concentration of TIMP's is increased and, consequently, the concentration of MMP's is decreased, there is a beneficial effect in terms of decreasing the degradation of collagen and its consequent accumulation, promoting a balance between the concentrations of MMPs and their tissue inhibitors and a reduction in the degree of vascular stiffening and, consequently, predisposition to cardiovascular events. Increased TIMP-1 may predict chronic heart failure in hypertensive patients. This study contradicts another study carried out by Dhingra et al. (2006) where the authors reported that an increase in TIMP-1 was associated with the incidence of hypertension and a greater risk of blood pressure progression (DHINGRA et al., 2009).

Other contradictory results have also been reported for TIMPs. For example, increased TIMP-1 has been reported in hypertensive vs. normotensive individuals (MUZAHIR et al., 2004; TAN et al., 2007). In contrast, other studies have reported unchanged (VISSCHER et al., 1994) or decreased TIMP-1 (KOREN et al., 2002).

Increased TIMP-1 in hypertension may represent a compensatory feedback mechanism in response to increased MMP, possibly favoring the accumulation of ECM and structural changes in cardiac and vascular tissue. Recent findings suggest that TIMPs play an important role in cardiovascular remodeling processes, independently of their MMP inhibitory activity, i.e. these inhibitors may play an important role in AH independently of the action of MMPs (TAYEBJEE et al., 2004).

It is known that the process of arterial stiffness becomes more pronounced as a result of aging or diseases associated with the cardiovascular system, such as AH, two characteristics presented by the patients under study.

No studies have been found in the literature that show a correlation between age and plasma levels of TIMP-1 in HTRNC, an aspect demonstrated in the results of this study.

According to the I Brazilian Positioning on Resistant Arterial Hypertension (2012), obesity is a predominant characteristic of patients with ARH. This study showed that the individuals in the two groups evaluated (HTRC and HTRNC) suffered from grade 1 obesity, as their BMI was $\geq$ 30mg/Kg .[2]

According to Cao (2007), MMPs, especially MMP-9 and MMP-2 and their tissue inhibitors (TIMP-1 and TIMP-2, respectively) play a crucial role in activating and inhibiting adipogenesis by regulating angiogenesis, They show a positive correlation with obesity, i.e. increased fat mass promotes greater production and release of these MMPs, contradicting the results observed in this study, which show a negative correlation between MMP-9 and BMI in CRH patients.

In addition to being diagnosed with RHT, the patients evaluated in this study had other associated pathologies such as type 1 diabetes and type 2 diabetes. A positive correlation between glycemia and TIMP-1 was observed in this study. Maxwell et al. (2009) reported a significant increase in TIMP-1 plasma levels (P<0.001) in type 1 diabetic patients compared to the non-diabetic control group. Lee et al. (2005) observed a significant increase in TIMP-1 levels (P<0.05) in type 2 diabetic patients compared to the control group. No studies were found in the literature showing a correlation between uric acid and TIMP-1 in HTRNC, as demonstrated in this study. The same is true for the lipid and renal profile of hypertensive patients, since there are no studies showing an influence exerted by MMP-9 and TIMP-1 on the plasma levels of the lipid fraction and renal markers in these patients.

With regard to aldosterone, hyperaldosteronism is an independent factor in arterial hypertension and consequently in the process of arterial stiffness, which may explain the negative correlation between plasma levels of aldosterone and biomarkers presented in this study.

Regarding plasma renin activity, the literature lacks information on the correlation with MMP-9 in HTRNC patients.

Another hypothesis initially raised was that circulating levels of arterial stiffness biomarkers would be positively correlated with PWV and PP, which was contradictory to the findings of this study that showed no correlation between these parameters and the biomarkers.

Pulse Wave Velocity (PWV) is widely used as an index of arterial stiffness and elasticity (FAVERO, 2009), and the properties of the arterial wall such as its thickness and the diameter of the arterial lumen are the factors that most influence PWV (SAFAR *et al.*, 2002).

It is known that the association between the MMP-9/TIMP-1 system and arterial stiffness is not simple and the existing data is very limited.

A study by Tan et al. (2007) showed that plasma levels of MMP-9 and TIMP-1 were significantly and positively correlated with OPV.

Elevated PWV and increased arterial stiffness are not only related to the stretching effect of high blood pressure, but also to early changes and abnormalities of the arterial wall in hypertensive disease (ASMAR, 2001).

These vascular alterations may be due to excessive degradation of the ECM in the arterial wall by the action of MMP-9, as shown in some studies described above.

According to Martins et al. (2011), the stiffening of the great arteries seems to be an inevitable consequence of ageing, i.e. this process becomes more pronounced at older ages and, according to the authors, is the most important determinant of arterial stiffness.

In the patients in this study, the process of arterial stiffness may have been completed or lost, since these individuals are of advanced age, which is directly related to the increase in OPV and PP, especially in the HTRNC group. This may explain why there was no correlation between the biomarkers under study and OPV and PP.

It is important to emphasize that the studies shown in this topic are studies carried out on patients with essential hypertension. There are no studies in the literature comparing levels of MMP-9 and TIMP-1 in patients with ARH, nor in CRHT and CRHT groups.

Other limitations associated with this study are the lack of a control group of normotensive individuals, carelessness in the collection of biological material, sample size (lack of sample calculation) and specific influences of individual clinical characteristics presented by the individuals in this study.

6 CONCLUSION

A greater association between TIMP-1 and uncontrolled resistant hypertension was evidenced in this study. The lack of information in the literature on this subject does not allow us to make conclusive statements, but we can suggest that TIMP-1 is an important element in the mechanisms of development and control of ARH, and could in future be used as a clinical biomarker to assess the development of this condition, as well as being used as a therapeutic target for the search for new drugs to control ARH more effectively, with a view to improving the patient's quality of life.

REFERENCES

AHMED, S. H.; et al. Matrix Metalloproteinases/Tissue Inhibitors of Metalloproteinases: Relationship Between Changes in Proteolytic Determinants of Matrix Composition and Structural, Functional and Clinical Manifestations of Hypertensive Heart Disease. **Journal of the American Heart Association,** v. 113, p. 2089-2096, 2006.

ANDRADE, A. C.; PEYNEAU, A. K. M. S.; ORNELAS, N. C. B. F. **Proposta de implantaçao de um Atença o Farmacèutica ao usuàrio hipertenso do Programa Farmâcia Popular do Brasil** - Faculdade de Ciências da Saùde da Universidade de Brasilia, Brasilia 2006.

ASMAR, R.; et al. Pulse wave velocity as endpoint in largescale intervention trial. The Complior® Study. Scientific, Quality Control, Coordination and Investigation Committees of the Complior Study. **Journal of Hypertension,** v. 4, p. 813-818, 2001.

BARRETO-FILHO, J. A. S.; KRIEGER, J. E. Genetics and hypertension: knowledge applied to clinical practice? **Revista da Sociedade de Cardiologia do Estado de Sào Paulo.** Sao Paulo,. v. 13, n. 1, p. 46-55, 2003.

BRAMLAGE, P. Hypertension in Overweight and Obese Primary Care Patients Is Highly Prevalent and Poorly Controlled. **American Journal of Hypertension.** v. 17, p. 904-910, 2004.

BRAVO, E. L. Phenylpropanolamine and other over-the-counter vasoactive compounds. **Journal of American heart association,** 1998.

BREW, K.; DINAKARPANDIAN, D.; NAGASE, H. Tissue inhibitors of metalloproteinases: evolution, structure and function. **Biochim Biophys Acta.** v. 1477, p. 267-83, 2000.

BRITTON, A., MCKEE, M. The relation between alcohol and cardiovascular disease in Eastern Europe: explaining the paradox. **Journal of Epidemiology Community Health.** v. 54, n. 5, p. 382-332, 2000.

CALHOUN, D. A. et al. Resistant hypertension: diagnosis, evaluation, and treatment: a scientific statement from the American Heart Association Professional Education Committee of the Council for High Blood Pressure Research. **Circulation.** v. 117, n. 25, p. 510-526, 2008.

CAO, Y. Angiogenesis modulates adipogenesis and obesity. **The Journal of Clinical Investigation.** v. 117, n. 9, 2007.

CECIL. **Tratado de medicina interna.** 21 ed. Rio de Janeiro: Guanabara Koogan. v. 2, 2001.

CHASE, A. J.; NEWBY, A. C. Regulation of matrix metalloproteinase (matrixin) genes in blood vessels: a multi-step recruitment model for pathological remodelling. **Journal of Vascular Research.** v. 40, p. 320-343, 2003.

CHOW, A. K.; CENA, J. S.; CHULZ, R. Acute actions and novel targets of matrix metalloproteinases in the heart and vasculature. **British Journal of Pharmacology.** v. 152, p. 189-205, 2007.

CIARONI, S. Hypertension artérielle résistante: quo vadis**? Journal Médica de Suisse.** v. 9, p. 607-612, 2005.

CUSHMAN, W. C. Alcohol consumption and hypertension. **Journal Clinic of Hypertension.** v. 3, n. 3, p. 166-170, 2001.

DHINGRA, R. et al. Relations of matrix remodeling biomarkers to blood pressure progression and incidence of hypertension in the community. **Circulation.** v. 119, p. 1101-1107, 2009.

FAVARELLI, M. H. C. **Involvement of the endothelium in the vascular response in patients with resistant arterial hypertension.** 2001. 138 p. - Campinas, 2001.

FAVERO, F. F. **Evaluation of arterial stiffness using the Pulse Wave Velocity (PWV) method in resistant and controlled hypertensive patients.** 2009, 88 p. - Campinas, 2009.

FERNANDEZ-PATRON, C.; RADOMSKI, M. W.; DAVIDGE, S. T. Vascular matrix metalloproteinase-2 cleaves big endothelin-1 yielding a novel vasoconstrictor. **Circulation Research.** v. 85, p. 906-911, 1999.

FONTANA, V. et al. Circilating matrixmetalloproteinase and their inhibitors in hypertension. **Clinica chimica ata.** v. 413, p. 656-662, 2012.

GIRIOLI, S. B. **Resistant Arterial Hypertension: Role of aldosterone and the effect of its antagonist spironolactone on cardiovascular remodeling and endothelial function.** 2009. 207 p. - Campinas, 2009.

GUEDIS, A. G. et al. White coat hypertension and its diagnostic importance. **Revista brasileira de hipertensâo.** v. 15, p. 46-50, 2008.

GUIMARAES, D. A. et al. Inhibition of extracellular matrix metalloproteinases: a possible therapeutic strategy in arterial hypertension? **Revista Brasileira de Hipertensâo.** v. 17, n. 4, p. 226-230, 2010.

HEYMANS, S. et al. Increased cardiac expression of tissue inhibitor of metalloproteinase-1 and tissue inhibitor of metalloproteinase-2 is related to cardiac fibrosis and dysfunction in the chronic pressure-overloaded human heart. **Circulation.** v. 112, p. 1136-1144, 2005.

INOUE, N. et al. Effect of angiotensin-converting enzyme inhibitor on matrix metalloproteinase-9 activity in patients with Kawasaki disease. **Clinica Chimica Acta.** v. 411, p. 267-269, 2010.

INTEGAN, H.D.; SCHIFFRIN, E. L. Vascular remodeling in hypertension: roles of apoptosis, inflammation, and fibrosis. **Hypertension.** v. 38, p. 581-587, 2001.

I Brazilian Positioning on Resistant Arterial Hypertension. Department of Hypertension of the Brazilian Society of Cardiology. Brazilian Archive of Cardiology. v. 99, p. 576-585, 2012.

III BRAZILIAN CONSENSUS ON ARTERIAL HYPERTENSION. **Hypertension - Diagnosis and Classification**. Brazil, 2001. chap. 1, p. 1-38.

KELLY, D et al. Plasma matrix metalloproteinase-9 and left ventricular remodelling after acute myocardial infarction in man: a prospective cohort study. **European Heart Journal.** v. 28, p. 711-718, 2007.

KELLY, D. et al. Plasma tissue inhibitor of metalloproteinase-1 and matrix metalloproteinase-9: novel indicators of left ventricular remodelling and prognosis after acute myocardial infarction. **European Heart Journal.** v. 29, p. 2116-2124, 2008.

KOREN, S. et al. Increased expression of matrix metalloproteinase-2: a diagnostic marker but not prognostic marker of papillary thyroid carcinoma. **Israel Medical Association** v 4, p. 247-51, 2004.

KRIEGER, E. M.; IRIGOYEN, M. C.; KRIEGER, J. E. Physiopathology of Hypertension. **Revista Sociedade de Cardiologia do Estado de Sâo Paulo**. Sâo Paulo, v. 9, n. 1, p. 1-7, 1999.

KUPAI, K. et al. Matrix metalloproteinase activity assays: Importance of zymography. **Journal of Pharmacological and Toxicological Methods**. v. 61, p. 205-209, 2010.

LESSA, I. Social impact of non-adherence to hypertension treatment

Arterial. **Revista Braeiloira Hipertensâo**, v. 13, n. 1, p. 39-46, 2006.

LINDSAY, M. M.; MAXWELL, P.; DUNN, F. G. **TIMP-1**: A Marker of Left Ventricular Diastolic Dysfunction and Fibrosis in Hypertension. **Journal of American Hearth Association.** v. 40, p. 136-141, 2002.

MANO, R. **Cardiology Manuals.** Year 6, 2003. **Available at:**

<www.manuaisdecardiologia.med.br>. **Accessed on**: 01/09/2012.

MARTINS, L. C. et al. Characteristics of resistant hypertension: aging, body mass index, hyperaldosteronism, cardiac hypertrophy and vascular stiffness. **Journal of Human Hypertension.** v. 25, p. 532-538, 2011.

MARTINS, L. M. B. **Variability of autonomic function in patients with resistant arterial hypertension.** 2010. 85 p. - Campinas, 2010.

MASSARO, M. et al. Statinsinhibit cyclooxygenase-2and matrix metalloproteinase-9 in human endothelial cells: anti-angiogenic actions possibly contributing to plaquestability . **European Society ofCardiology** .

Cardiovascular Research. v.10. p. 375, 2009.

MURPHY, G.; NAGASE, H. Progress in matrixmetalloproteinase research. **National Institute of Health.** v. 29. n. 5. p. 290-308, 2008.

MUZAHIR, H. et al. Matrix Metalloproteinase-9 and Tissue Inhibitor of Metalloproteinase-1 in hypertension and their relantionship to cardiovascular risk and treatment. **American Journal of Hypertension.** v. 17, p. 764-769, 2004.

MUZAHIR, H. et al. Tissue inhibitor of metalloproteinase-1 and matrix metalloproteinase-9 levels in patients with hypertension. **American Journal of Hypertension.** v. 17, p. 770-774, 2004.

NAGASE, H.; VISSE, R.; MURPHY, G. Structure and function of matrix metalloproteinases and TIMPs. **Cardiovascular Research.** v. 69, p. 562 - 573, 2006.

National Institute of Health State-of-The Science Conference Statement: **Tobacco use: Prevention, Cessation, and Control.** NIH Conference. Ann Intern Med. p. 145-839, 2006.

ONAL, I. K. et al. Serum levels of MMP-9 and TIMP-1 in primary hypertension and effect of antihypertensive treatment. **European Journal of Internal Medicine.** v. 20, p. 369-372, 2009.

PHILIPPE, E. et al. Biochemistry and Molecular Biology of Gelatinase B or Matrix Metalloproteinase-9 (MMP-9). **Critical Reviews in Biochemistry and Molecular Biology.** v. 37, p. 375-536, 2002.

PIMENTA, E. M. D. et al. Mechanisms and treatment of resistant Hypertension. **The Journal** of Clinical Hypertension. v. 10, n. 3, 2008.

REINHARDT, D. et al. Cardiac remodelling in end stage heart failure: upregulation of matrix metalloproteinase (MMP) irrespective of the underlying disease, and evidence for a direct inhibitory effect of ACE inhibitors on MMP. **Basic Research Heart,** 2002.

RENOVATO, R. D.; TRINDADE, M. F. Pharmaceutical Care for Hypertension in a Pharmacy in Dourados, Mato Grosso Do Sul. **Infarma.** Mato Grosso do Sul, v.16, n. 11-12, p. 49-55, 2004.

RODRIGUES, C. I. S.; ALMEIDA, F. A. Refractory arterial hypertension: an overview. **Revista Brasileira de Hipertensâo.** v. 11, n. 4, p. 218-222, 2004.

ROSARIO, T. M. et al. Prevalence, Control and Treatment of Systemic Arterial Hypertension in Nobres - MT. **Arquivo Brasileiro de Cardiologia.** v. 93, n. 6, p. 672-678, 2009.

SAFAR, H et al. Aortic pulse wave velocity, an independent marker of cardiovascular risk. **Arch Mal Couer Vaiss.** v. 21, p. 215-1218, 2002.

SCUOTTO, F. et al. True resistant hypertension: knowing how to identify and manage it. **Revista Brasileira de Hipertensâo.** v. 16, p. 134-138, 2009.

SILVA, J. L. L.; SOUZA, S. L. Risk factors for systemic arterial hypertension versus teaching lifestyle. **Revista Eletrônica de Enfermagem.** v. 6, n. 3, p. 330-335, 2004.

SLUIJTER, J. P. G. et al. Vascular remodeling and protease inhibition - bench to bedside. **Cardiovascular Research.** v. 69, p 595- 603, 2006.

SOUZA, L. A. **Functional evaluation of venous and arterial endothelium in patients with refractory arterial hypertension.** 2009. 203 p. - Campinas - Brazil, 2009.

SOLINI, A. et al. Family history of hypertension, anthropometric parameters and markers of early atherosclerosis in young healthy individuals. **Journal of Human Hypertension.** v. 23, p. 801-807, 2009.

SORBI, D. et al. Captopril inhibits the 72 kDa and 92 kDa matrix metalloproteinases. v. 44, p. 1266-72, 1993.

TAN, J. et al. Impact of the metalloproteinase-9/tissue inhibitor of metalloproteinase-1 system on large arterial stiffness in patients with essential hypertension. **Hypertension Research.** v. 30, n. 10, 2007.

TANYA, L. et al. Matrix Metalloproteinase-9 Genotype Influences Large Artery Stiffness Through Effects on Aortic Gene and Protein Expression. **Journal of the American Heart**

Association. v. 24, p. 1479-1484, 2004.

THADHANI, R.; CAMARGO, C. A. Jr.; STAMPFER, M. J.; CURHAN, G. C.; WILLETT, W. C.; RIMM, E. B. Prospective study of moderate alcohol consumption and risk of hypertension in young women. **Archiev International of Medicine.** v. 162, n. 5, p. 569-574, 2002.

TOLEDO, J. C. Y. **Refractory arterial hypertension: phenotypic characterization, cardiovascular morpho-functional evaluation and correlation with genetic polymorphisms of the renin-angiotensin system and endothelial nitric oxide synthesis.** 2006. 211 p. - Campinas, 2006.

VELAGALETI, R. S. et al. Relations of Biomarkers of Extracellular Matrix Remodeling to Incident Cardiovascular Events and Mortality. **Journal of the American Heart Association.** v. 30, p. 2283-2288, 2010.

VELGALETI, R. S. et al. Relations of biomarkers of extracellular matrix remodeling to incident cardiovascular events and mortality. **Arteriosclerosis and Thrombosis Vascular of Biology.** v. 30, p. 2283-2288, 2010.

VI BRAZILIAN GUIDELINES FOR ARTERIAL HYPERTENSION. Brazilian Society of Cardiology. Brazil. Brazilian Archive of Cardiology. v. 95, p. 151, 2010.

VISSCHER, D. W. et al. Enhanced expression of tissue inhibitor of metalloproteinase-2 (TIMP-2) in the stroma of breast carcinomas correlates with tumor recurrence. **international Journal of Cancer.** v. 59, p. 339-44, 1994.

VISSE, R.; NAGASE, H. Matrix Metalloproteinases and Tissue Inhibitors of Metalloproteinases: Structure, Function, and Biochemistry. **Journal of the American Heart Association.** v. 92, p. 827-839, 2003.

YASMIN, S. W. et al. Matrix Metalloproteinase-9 (MMP-9), MMP-2, and Serum Elastase Activity Are Associated With Systolic Hypertension and Arterial Stiffness. **Journal of American Heart Association.** v. 25, p. 372-378, 2005.

ZHOU, S. Matrix metalloproteinase-9 polymorphism contributes to blood pressure and arterial stiffness in essential hypertension. **Journal of Human Hypertension.** v. 21, p. 861-867, 2007.

Printed by Books on Demand GmbH, Norderstedt / Germany